Understanding Ecosystems

Lesson 1

What Are the Parts of an Ecosystem? 2

Lesson 2

What Factors Influence Ecosystems? 8

Lesson 3

How Do Humans Affect Ecosystems? 16

SCHOOL PUBLISHERS

Orlando   Austin   New York   San Diego   Toronto   London

Visit *The Learning Site!*
www.harcourtschool.com

Lesson 1

VOCABULARY
environment
ecosystem
population
community

What Are the Parts of an Ecosystem?

An **environment** is all of the living and nonliving things that affect an organism.

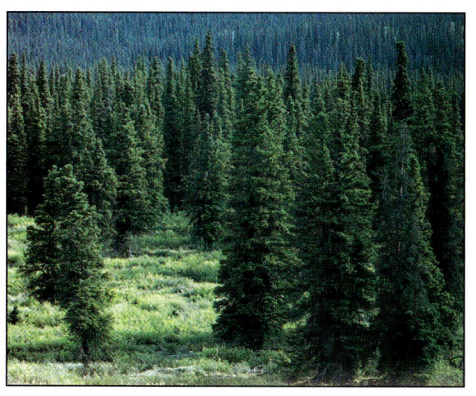

An **ecosystem** is all the living and nonliving things in one place. An ecosystem can be as small as the space under a rock or as large as a forest.

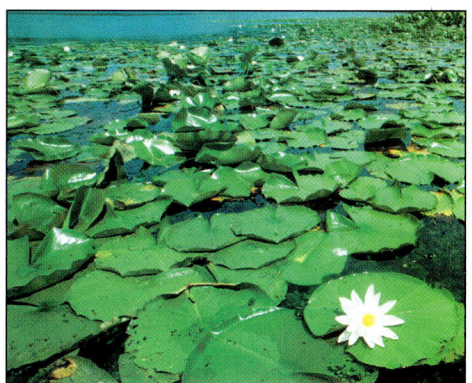

A **population** is a group of all of the same kind of living things living in the same ecosystem. These water lilies are part of one population.

A **community** is made up of all of the populations of living things that live in the same place.

READING FOCUS SKILL
MAIN IDEA AND DETAILS
The main idea is what the text is mostly about. Details tell more about the main idea.
Look for details that tell about an ecosystem.

Ecosystems

An **environment** is all the living and nonliving things that affect a living thing. The living things in your environment are people, animals, and plants. The nonliving things are water, air, soil, and weather.

An **ecosystem** is all the living and nonliving things in a place. It also has a climate. Climate is the average weather over many years. It includes temperature and rainfall.

▼ Prairie smoke plants

▼ Prairie dogs

Ecosystems may be different sizes. A small ecosystem might be the space under a rock. Insects may live there. Nonliving things include soil, water, and air.

A large ecosystem might be a forest. This ecosystem is home to many plants and animals. Each living thing finds what it needs in the forest. A forest also has nonliving parts. They include water, air, soil, and climate.

Focus Skill **Name the two parts of every ecosystem and give two examples of each part.**

Moose live in a forest ecosystem. ▶

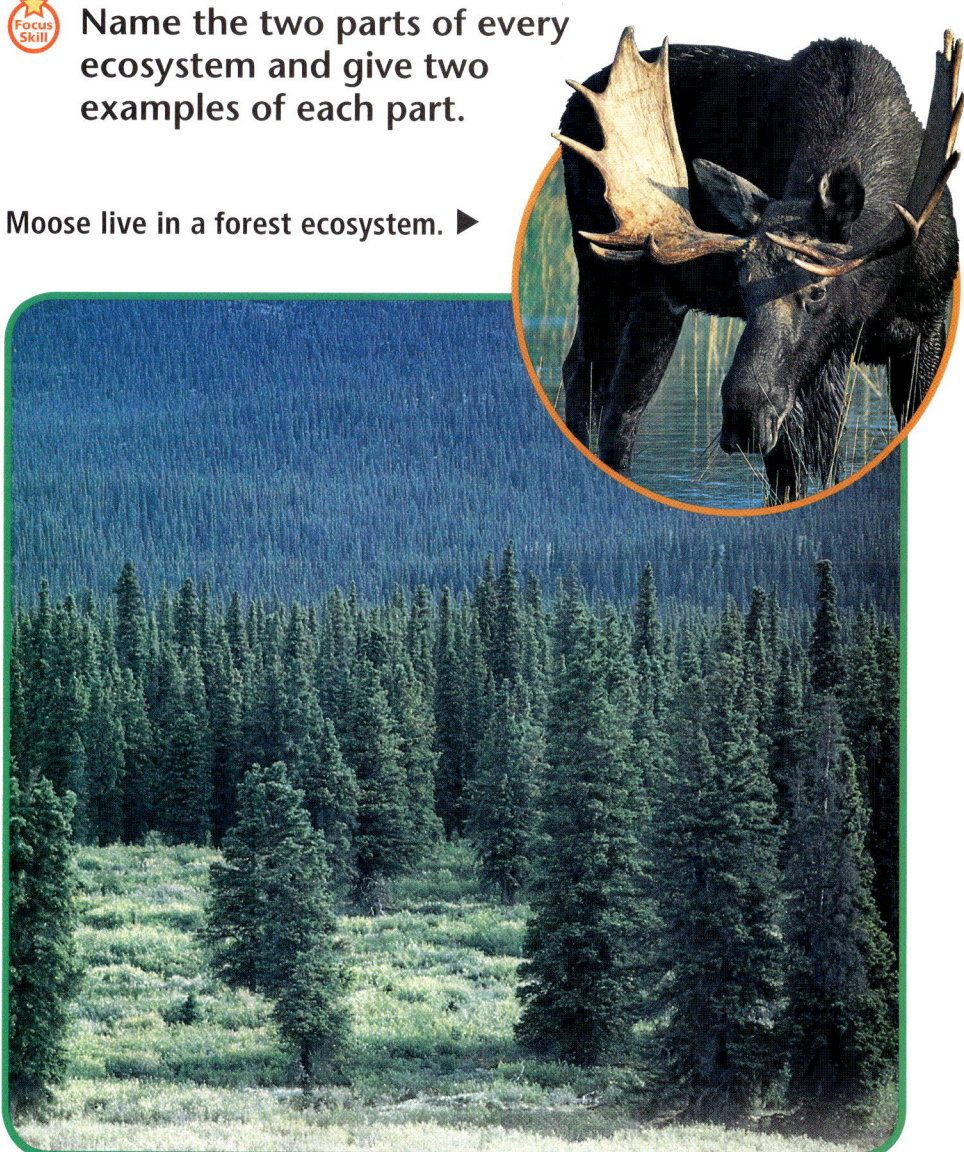

5

Individuals and Populations

One plant is an individual. One animal is an individual, too. A bee or pine tree is an individual.

A **population** is a group of the same kind of individuals living in the same ecosystem. Examples of populations include a hive of bees or a group of pine trees. Members of some populations, like frogs, live alone. Other populations, like wolves, live in groups.

Some populations, like blackbirds, can live in several ecosystems. If one ecosystem cannot meet their needs, they can move.

 Name another example of an individual and a population.

Blackbird population ▼

Communities

A <mark>community</mark> is all the populations that live in the same place. It includes all the plants and animals that live there.

The plants and animals in a community depend on each other. Some animals eat the plants. Other animals eat the plant eaters. Animals help plants by spreading seeds. Plants provide places for animals to live and hide.

 How do plants and animals in a community depend on each other?

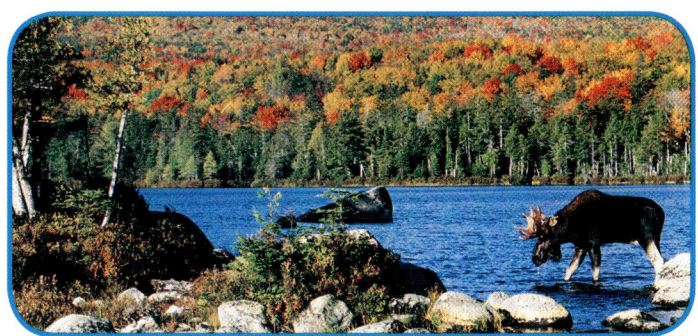

◀ Many populations make up this community.

Review

Complete this main idea statement.

1. A forest _____ is made up of both living and nonliving things.

Complete these detail statements.

2. A rabbit is an example of an _____.

3. A group of robins living in a forest is called a _____.

4. A _____ includes all the plants and animals that live in the same place.

Lesson 2

VOCABULARY
biotic
abiotic
diversity

What Factors Influence Ecosystems?

The living parts of an ecosystem are **biotic**. Plants and animals are biotic factors in an ecosystem. The nonliving parts of an ecosystem are **abiotic**. Air, water, sunlight, and soil are abiotic factors in an ecosystem.

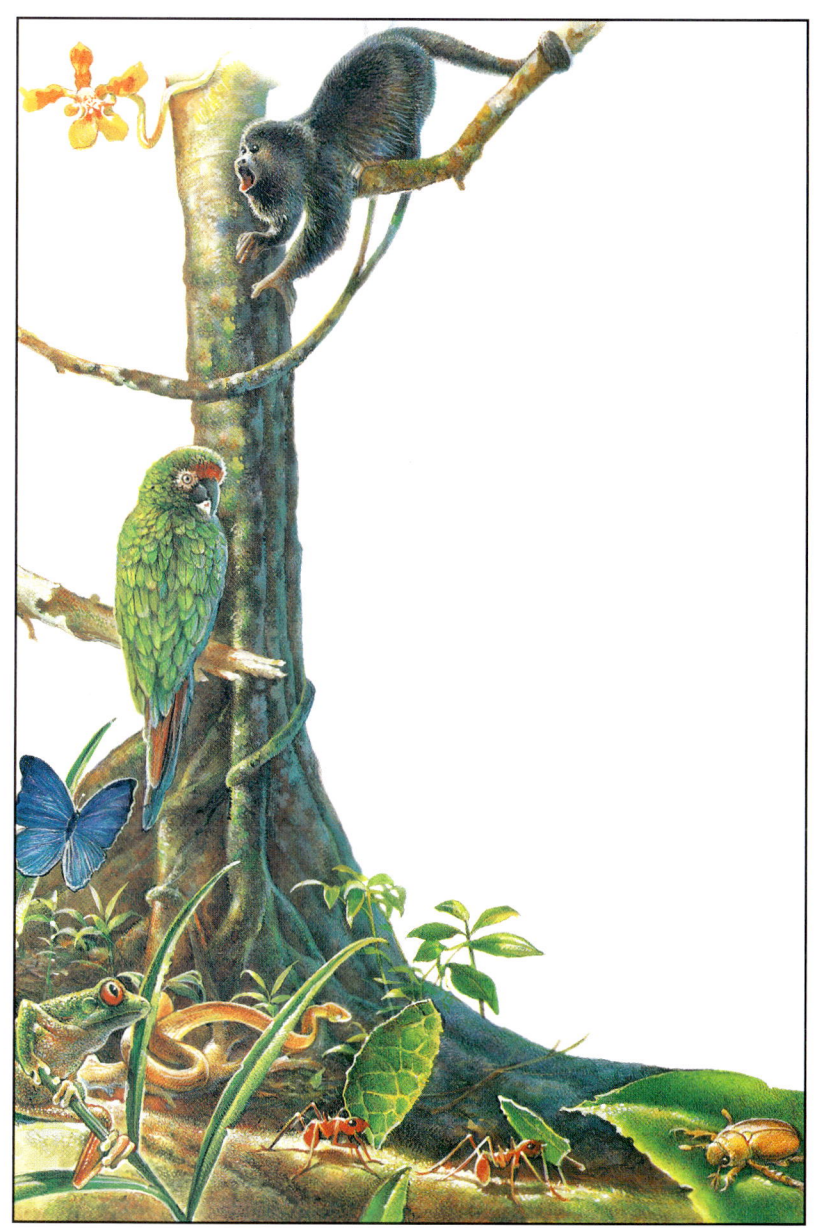

Diversity is the variety of living things. A rain forest has the most diversity of all ecosystems.

READING FOCUS SKILL
CAUSE AND EFFECT

A cause is what makes something happen. An effect is what happens.

Look for the effects that living and nonliving things have on ecosystems.

Living Things Affect Ecosystems

Plants and animals are biotic factors in an ecosystem. Biotic factors are the living parts of an ecosystem. These factors affect the ecosystem. They also affect each other in many ways.

Animals use plants in many ways. Plants provide food and shelter.

Animals can help plants. By eating plants, they give more space for other plants.

Animals can also harm plants. Too many plant eaters can wipe out plants.

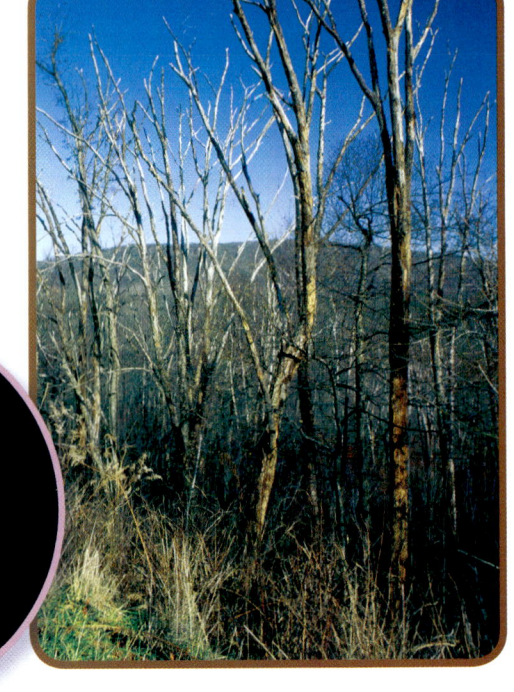

Gypsy moths stripped these trees. ▼

Animals can affect one another. More wolves can lead to more rabbits being eaten. Fewer rabbits may cause hungry wolves. Fewer rabbits may cause more plants to grow.

A change in plants can cause a change in animals. If dry weather kills plants, rabbits have less to eat. Fewer rabbits causes wolves to go hungry.

Sometimes new plants or animals can change an ecosystem. New plants can crowd out other plants. New animals may wipe out other animals. Or they may eat all the plants.

Disease can change an ecosystem. It can kill plants or animals. This can change an ecosystem's balance.

Focus Skill **Tell how an increase in plants could affect an ecosystem.**

Too many deer can wipe out plants. ▶

Nonliving Things Affect Ecosystems

Air, water, soil, and sunlight are the abiotic parts of an ecosystem. **Abiotic** factors are the nonliving parts of an ecosystem. They are as important as the living things in an ecosystem.

A change in water can affect all living things in an ecosystem. Too little rain causes plants to die. Without plants, animals may die or leave.

Soil also affects an ecosystem. Rich soil means many plants will grow. Poor soil leads to fewer plants. Without plants, fewer animals can live in the ecosystem.

Air also affects living things. Bad air can harm many living things.

 Tell how a change in the water supply could affect a rabbit.

Nonliving factors ▶

12

Climate Affects Ecosystems

Climate is an abiotic factor. Climate includes the amount of rainfall and sunlight. It also includes air temperature. This map shows the main climates of Earth.

Climate affects the soil. Some climates allow many living things to grow.

Climate also affects the kinds of living things in an ecosystem. Some living things need mild or wet climates. Others are suited to cold or dry climates.

 Tell what would happen to an ecosystem if its climate changed.

Main climate zones of Earth ▼

14

Diversity

Some ecosystems have only a few living things. Others have many. Diversity is the variety of living things.

Rain forests have the most diversity of all Earth's ecosystems. This is because they provide the things that living things need.

Other ecosystems have little diversity. In these ecosystems, living things may find it hard to survive.

 Tell what leads to a diversity of living things in an ecosystem.

Rain forest ecosystem ▶

Review

 Complete these cause and effect statements.

1. A decrease in wolves can cause an _____ in rabbits.
2. Poor soil may lead to _____ plants.
3. Too little rain can cause plants and animals to _____.
4. The _____ of plants and animals in an ecosystem is a result of how many living things can find what they need to live.

Lesson 3

VOCABULARY
pollution
habitat
  restoration

How Do Humans Affect Ecosystems?

Harmful waste products that enter an ecosystem are **pollution**. Pollution may harm water, land, or air.

Habitat restoration happens when people return a natural environment to its original condition. Planting trees is one way to repair damage to an environment.

READING FOCUS SKILL
COMPARE AND CONTRAST
When you **compare and contrast**, you tell how things are alike and different.
Compare positive and negative effects of changes made by humans on ecosystems.

Humans Within Ecosystems

Natural resources are parts of an ecosystem that people use. They include air, water, soil, trees, and minerals.

Some resources, such as air, water, and trees, can be replaced. Others cannot. Resources that cannot be replaced include coal, oil, and iron. Once supplies are used, these resources will be gone.

Lakes, fresh air, and sunlight are natural resources. ▼

People use resources in many ways. They use wood to build homes. They use sand to make glass. They use iron to make steel.

People raise crops to feed themselves. They also raise crops to make medicines.

 How are coal and water alike and different?

▲ Medicine is made from this tree.

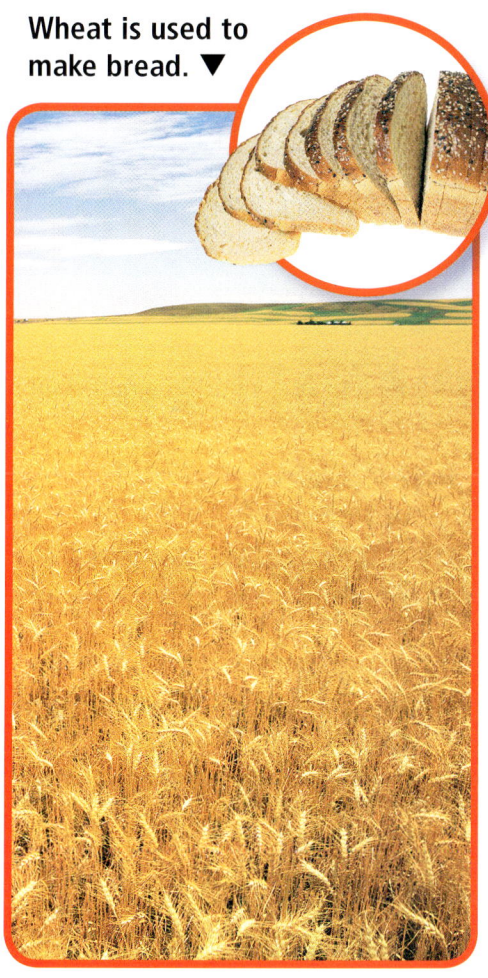

Wheat is used to make bread. ▼

Negative and Positive Changes

People's actions can harm an ecosystem. People may clear land. This can destroy habitats. Animals can no longer meet their basic needs. They must move or die.

People can cause pollution. **Pollution** happens when harmful materials mix with water, air, or soil.

Water pollution can happen if chemicals wash into water. Trash can also pollute water. Soil pollution can come from trash and chemicals.

Air pollution can come from cars. Smoke from smokestacks can also pollute the air. Some chemicals form acid rain. This harms plants and pollutes lakes.

◄ Cars can cause air pollution.

Tree planting ▶

People help ecosystems by habitat restoration. **Habitat restoration** happens when people return a natural environment to the way it was. Planting trees can do this. Making wetlands can, too.

People also help when they pollute less. They do this by reducing harmful gases. Recycling and using fewer chemicals also helps ecosystems.

 Tell how people can harm and help an ecosystem.

▼ Bicycles don't release pollution.

21

Planning for Change

Earth's human population keeps growing. More people need more homes, roads, and factories. Building can harm ecosystems.

Before building, people must plan. They must think about biotic factors in an ecosystem. Filling wetlands can harm plants and animals. Washing soil into a river can harm fish and plants.

Planners draw plans and build models. ▼

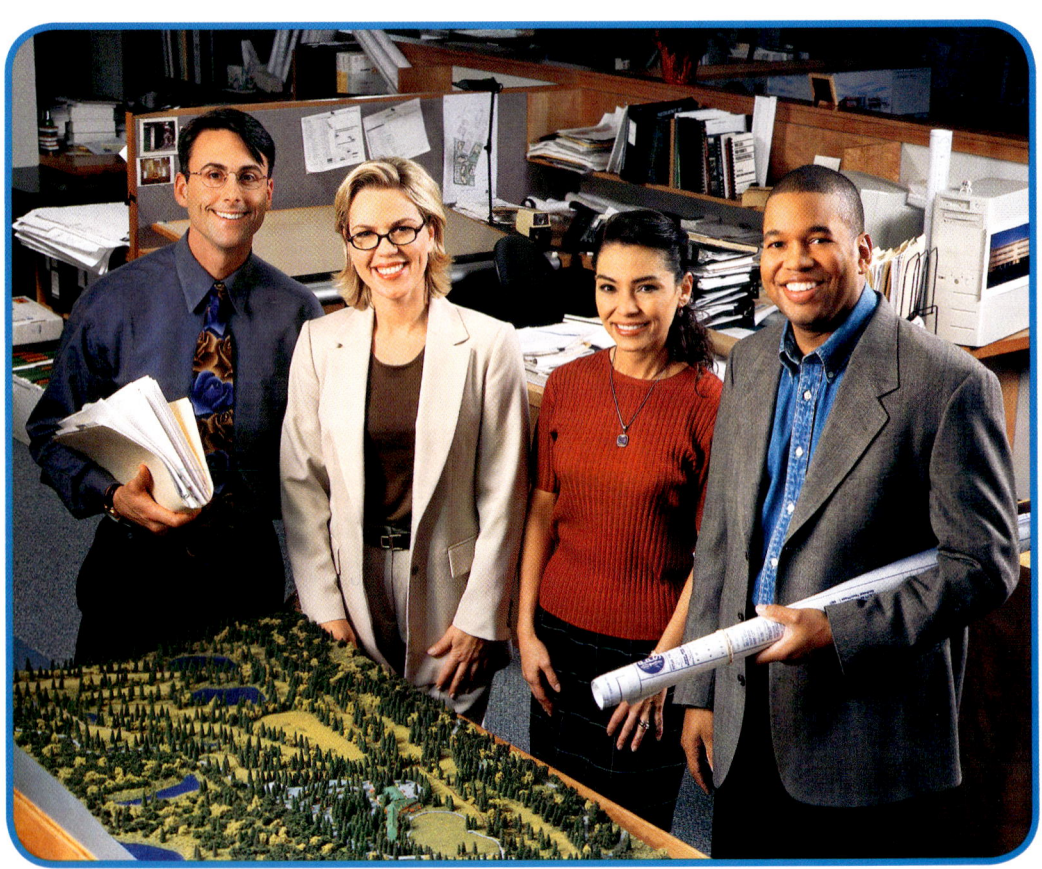

Planners must also think about abiotic factors. This includes climate, rainfall, and type of soil.

Careful planning helps protect the community and physical environment of an ecosystem.

 Tell about the different kinds of factors people must think about when building.

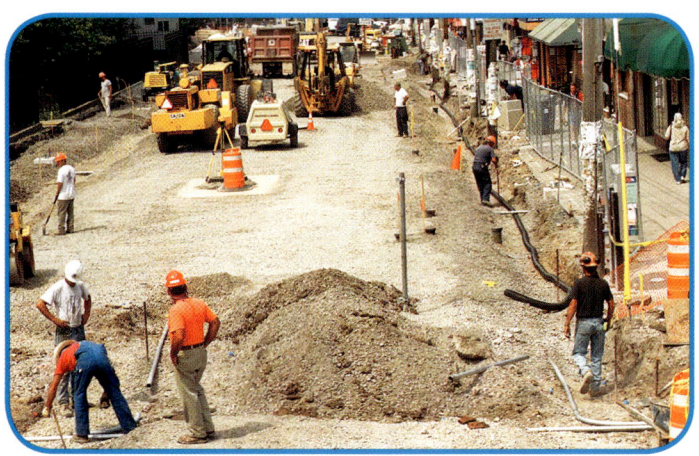

Construction site ▲

Review

Complete these compare and contrast statements.

1. Oil, wheat, air, and iron are all _____.
2. Trees, water, and air are resources that can be _____, while oil and coal cannot.
3. Both recycling and reducing the use of chemicals can help to reduce _____.
4. Planting trees and creating wetlands are both examples of _____.

GLOSSARY

abiotic (ay•by•AHT•ik) of the nonliving parts of an ecosystem

biotic (by•AHT•ik) of the living parts of an ecosystem

community (kuh•MYOO•nuh•tee) all the populations of organisms living together in an environment

diversity (duh•VER•suh•tee) a great variety of living things

ecosystem (EE•koh•sis•tuhm) all the living and nonliving things in one place

environment (en•VY•ruhn•muhnt) all of the living and nonliving things that affect an organism

habitat restoration (HAB•i•tat res•tuh•RAY•shuhn) returning a natural environment to its original condition

pollution (puh•LOO•shuhn) waste products that damage an ecosystem

population (pahp•yuh•LAY•shuhn) all the individuals of the same kind living in the same environment